RAGGEDY ANN & ANDY COLLECTIBLES

A Handbook and Price Guide

Jan Lindenberger

77 Lower Valley Road, Atglen, PA 19310

A variety of Raggedy Ann and Andy dolls being held by (left to right) Ashley Lindenberger, Alysia and Jessica Knight, the author's granddaughters.

Printed in Hong Kong
ISBN: 0-88740-782-X

Book Design by Audrey L. Whiteside.

Library of Congress Cataloging-in-Publication Data

Lindenberger, Jan.
    Raggedy Ann and Andy Collectibles: a handbook and price guide/Jan Lindenberger.
        p. cm.
    ISBN: 0-88740-782-X (paper)
    1. Raggedy Ann and Andy Dolls--Collectibles—Catalogs.
I. Title
NK4894.3.R34L56 1995
688.7'221'0973075—dc20   95-19703
                                CIP

Published by Schiffer Publishing, Ltd.
77 Lower Valley Road
Atglen, PA 19310
Please write for a free catalog.
This book may be purchased from the publisher.
Please include $2.95 postage.
Try your bookstore first.

We are interested in hearing from authors
with book ideas on related subjects.

Contents

Acknowledgements

A very special thank you to Annie and Jim Luke for encouraging me to write this book to its completion. They opened their home to me and allowed me to arrange and rearrange it in order to photograph their wonderful collection.

Joel Martone gave me much of the valuable information for this book and allowed me to photograph his extensive Raggedy Ann and Andy collection. Joel has been an avid collector of Raggedys, dolls, teddy bears, toys and Christmas collectibles for 25 years. He is a prolific creator of original hand-made Raggedy Sam and Sue's, bears, bunnies, dolls and marionettes, and is a long time member of the Original Artists Council of America. He is also co-owner of a shop called "Rhyme or Reason" in Colorado Springs, Colorado. Some of his original work appears on later pages in this book. Thank you Joel.

Annette Mahante from Broomfield, Colorado, deserves recognition for her hospitality and kindness and also for sharing her vast and varied collection.

Another special thank you to Gwen Daniels, an antique dealer from O'Fallon, Missouri, who recently had a part in *The Raggedy Ann and Andy Family Album*. Her knowledge and the photographs of her quality collection of rare and one of a kind Raggedy Anns and Andys was certainly a needed blessing.

I also thank the following people who allowed me to photograph in their shops or homes, and anyone else I may have forgotten.

Evelyn Haling, Canon City, Colorado
Conestoga Antique Mall, Arvada, Colorado
Antique Gallery, Colorado Springs, Colorado
Antique Mart, Colorado Springs, Colorado
Nevada Avenue Antique Mall, Colorado Springs, Colorado
Sara Troup, Colorado Springs, Colorado
Stage Stop Antique Mall, Denver, Colorado
Stuff Antique Mall, Denver, Colorado
Gilleys Antique Mall, Plainfield, Indiana
Quonset Hut Antique Market, Adamstown, Pennsylvania

Introduction

Raggedy Ann and Andy continue to be among the most treasured of all dolls. Children everywhere have fallen in love with these rag dolls and cherished them through adulthood.

Johnny Gruelle created Raggedy Ann when his only daughter, Marcella, found an old rag doll in her grandmother's attic. The doll's face was faded, but Gruelle, an illustrator and cartoonist, painted a new one with a big smile. Marcella suffered from a long-term illness and loved to spend hours playing with her dolls. Gruelle named the doll after two poems by his friend James Whitcomb Riley, "The Raggedy Man" and "Orphan Annie." The character of Raggedy Ann and her adventures was at the center of a series of stories he made up to entertain his frail daughter. Raggedy Ann became Marcella's constant companion and treasured friend.

After Marcella died in 1916, Gruelle started to write down the stories he had invented during her illness. Marcella's room and toy's became the inspiration for many of these books, with Raggedy as the centerpiece. Eventually he wrote and illustrated 25 books in the series.

Prompted by public interest in his Raggedy Ann series of books, in 1918 Gruelle and his family made several dozen original Raggedy Ann dolls to sell as storybook companions. Later that year, Gruelle licensed the P.F. Volland Company, which had published the books, to manufacture the dolls based on the Gruelle's 1915 patent design. Since then the design for the dolls has been licensed to several different manufacturers, with subtle variations in facial features and dress showing up over the years.

Shortly after the dolls were first produced, Gruelle received an unexpected package. A note explained that the owner had been a childhood friend of Gruelle's mother; their mothers had made a pair of girl and boy dolls for the two playmates. Now the woman had recognized Raggedy Ann as her friend's doll and wanted to reunite the pair. In the package was another old rag doll: Raggedy Ann's "twin brother" Andy.

With this inspiration, the Gruelle family started licensing production of other characters from the books, and also expanded into other types of toys. Beloved Belindy and the Camel with the Wrinkled Knees soon joined Raggedy Ann and Andy. Coloring books, puzzles and games, and other accessories were also manufactured. All of these things are avidly collected today.

The original Raggedy Ann dolls had an all-cloth body, shoe button eyes, a painted face, brown yarn hair, a dress, pantaloons, a pinafore, stripped legs, and black cloth shoes. Over the years, many variations on this basic theme have added to the doll's charm and universal appeal, and increased the possi-

5

bilities for collectors. The dolls have been manufactured with red yarn hair instead of brown, printed rather than painted faces, different expressions (always retaining that loveable smile, however), and many styles of dress. In addition, a musical Ann and Andy were made in the 1970s, and the current manufacturer has released a Christmas edition, with Ann and Andy dressed in red and green. But the dolls' timeless message of love, symbolized by their red hearts with its simple imprint, has never changed.

It is believed the dolls produced by the P.F. Volland Company, the original manufacturer, had a real candy heart inside them. After complaints from parents that the dolls got sticky when children sucked on them for the candy taste, the candy heart was replaced with a cardboard one. In 1935, Mollye Goldman was the first to manufacture the dolls with a heart imprinted on Ann's chest. Since then, both Ann and Andy have had the famous "I Love You" hearts. Ann's and Andy's hearts have woven their way into our fondest childhood memories.

This collector's guide highlights some of the many variations of these dolls now available to collectors. It includes historical and pricing information for the avid collector, and plenty of photographs for those who simply enjoy the doll's timeless appeal. *Raggedy Ann and Andy and Their Friends* will be a welcome addition to those who cherish their Raggedy Ann and Raggedy Andy dolls...young and old, children and collectors alike.

Quadruplets.

This is only a general price guide. Prices may differ according to area location and/or condition. Also shop prices may differ from auction prices and flea market prices.

Manufacturers of Ann and Andy

P.F. Volland Company
1918-1934

In 1918, P.F. Volland Co. was a publisher of inspirational books and gifts. That September the first "Raggedy Ann stories" were printed. These stories were so successful that the company saw a probable market for dolls to accompany the books. Gruelle was approached with this idea and granted to Volland the exclusive right to manufacture the Raggedy Ann doll. Volland contracted with Non-Breakable Toy Co. for the manufacturing of these first commercially made dolls, using Gruelle's patent design.

In 1919, the president and owner of the company, Paul Volland was shot and killed. His successor was T.J. Clampitt, the secretary / treasure of the company.

Early in that same year, Gruelle had approached Clampitt with an idea for a Raggedy Andy book. The idea was accepted and in 1920 the finished manuscripts were presented and the publication of eleven Raggedy Andy books followed.

Gruelle decided at this point to take greater control over the manufacturing of the doll that would compliment the books. He contracted Beers, Keeler and Bowman for this endeavor. The dolls were approved by P.F. Volland Co. and in 1920 were produced to be sold, along with the books by the Volland Co.

In 1924 Gruelle wrote another book, his first in four years. It was a combination of Ann and Andy called *Raggedy Ann and Andy and the Camel with the Wrinkled Knees*. This book was also published by Volland Co.

In 1924 Volland merged with the Gerlach-Barklow Co. and moved from Chicago to Joliet, Ill. This concerned Gruelle, but the concern proved unfounded as Volland Co. continued to produce his books in the same good manner as before the merger.

Gruelle's contract called for him to write one new book a year, and he did. A book on Beloved Belindy, a maternal Afro-American mammy, was introduced in 1926, as was a Belindy doll.

Due to the depression conditions in the 1930s, Volland Co. stopped publication of Gruelle's books. In 1934 the Volland Co. assigned back to Gruelle, all control of the copyrights and trademarks for the Raggedy Ann and Andy books and dolls.

M.A. Donohue and Company
1934

Gruelle authorized reprinting of his Volland Raggedy Ann books by the M.A. Donohue Co. In a year they had published more than 30,000 copies and continued to do so.

Gruelle's wife Myrtle kept this going with licensing from her partnership, "Johnny Gruelle Co." in 1941.

Exposition Toy and Doll Manufacturing Company
1934-35

In 1934 Gruelle negotiated with Sam Drelich, who represented the Exposition Toy and Doll Manufacturing Company, to manufacture and sell a brand new Gruelle-authorized Raggedy Ann doll, to replace the now defunct Volland model.

Several thousand dolls were produced under Gruelle's trademark, but due to the arrival of Mollye Goldman and the battle that ensued, the Exposition Toy and Doll Manufacturing Co. never reaped any reward for their endeavor.

Whitman Company
1935

Gruelle contracted with this company to publish his new Raggedy Ann books. This company distributed its books principally through five-and-dime stores. This company published two books in 1935, *Raggedy Ann and the Left Handed Safety Pin* and *Raggedy Ann Cut Out Paper Dolls*. It sold over 100,000 copies of both books in a year.

One of Gruelle's best works, published by Whitman, was *Raggedy Ann in the Golden Meadow*. It was an oversized book based on his early newspaper serials from 1922. This book featured his finest illustrations, motifs and favorite characters.

Johnny Gruelle wrote and illustrated two other books for Whitman Publishing Company that were never published. They were titled *Raggedy Ann's Kindness* and *Raggedy Ann and the Queen*.

Molly-es Doll Outfitters
1935-1938

Mollye and Meyer Goldman started producing her version of the Raggedys in 1934. The dolls first showed up at the New York toy fair in 1935.

Mollye presumed that Gruelle's patent rights, having expired in 1929, gave her the right to manufacture a Raggedy Ann and Raggedy Andy doll.

In March of 1935 Mollye applied for her own patent for Raggedy Ann and Andy. She also tried to get a license from Gruelle to continue her manufacturing of the dolls, but he had already contracted the Exposition Toy and Doll Manufacturing Co., and refused her offer.

The legal court battles began and throughout the battles, Mollye continued to advertise, manufacture, and sell her dolls. The courts finally ruled in favor of Gruelle in December, 1937, but by that time Gruelle had been unable to interest anyone else in producing an authorized Raggedy doll. It was reported that Mollye had made over a million dollars on the Raggedy dolls she produced. Meanwhile Gruelle lost hundreds of thousands in court costs and lost sales.

In March of 1938, Meyer and Mollye Goldman filed in court again, hoping to get a reversal that would enable them to start the making the Raggedy dolls again. This was probably prompted by the fact that they had to file for bankruptcy. In April their filing was denied and Molly-es' Doll Outfitters was declared insolvent and its assets sold. Johnny Gruelle did not have to go through this last battle. He passed away in January of 1938. His wife, Myrtle, that took on the Raggedys.

Georgene Novelties Company
1938-1962

At the end of 1938, Myrtle Gruelle finalized an agreement granting this company the right to produce their own version of the Gruelle's Raggedys. This was a company specializing in character dolls. This agreement was for a period of 25 years, presumably to prevent another "Mollye Goldman" incident.

In the 1940s they added an asleep-awake Raggedy Ann, and later Beloved Belindy and Camel with the Wrinkled Knees.

Johnny Gruelle Company
1939-1943

In 1939 a partnership was formed between Myrtle Gruelle and Howard Cox (her son and former Volland associate). This was called the Johnny Gruelle Company. This partnership's reason for being was to oversee all copyrights, trademarks, publishing, licensing rights, etc. to Raggedy Ann and Andy and related characters. Using old Raggedy serialized stories they produced five Raggedy Ann books under this imprint.

During the next several years the Johnny Gruelle Company licensed dozens of consumer goods like toys, bedding and even a cartoon in 1940. These were added to the continuing licenses with Georgene Novelties for the dolls and M.A. Donohue and Company for the reprint books, both set up by Johnny Gruelle. In addition to these other licenses were added. These included the Volland Co.(for greeting cards), Holgate Brothers, Halsam Safety Block Company, Milton Bradley/McLoughlin Brothers, Dell Publishing and Saalfield Publishing Company.

In 1943 the Johnny Gruelle Company was dissolved. Howard Cox retained the right to use and license the Raggedy Ann and Andy characters and others in any form except stuffed. He was then the Johnny Gruelle Company.

Myrtle Gruelle still had the control of all the Raggedy Ann properties and rights.

Bobbs-Merrill Company
1960-1980

In 1960 the Bobbs-Merrill Company acquired the sole and exclusive right an a license to use and sublet others to use the Raggedy Ann and Andy characters. This let the Johnny Gruelle Company out and began a new marketing era for the Raggedys. They produced four new books from the Gruelle serial stories under the Bobbs-Merrill imprint.

In the 1970s Bobbs-Merrill company was acquired by ITT, and in the early 1980s the Bobbs-Merrill trade division was purchased by Macmillan Inc. With the right to assign book copyright and trademark rights of Raggedy Ann and Andy and grant merchandise licenses, they produced hand puppets, silk screened Raggedy dolls with music boxes, and a new line of grooming aids.

Knickerbocker Toy Company
1962-1982

At the end of a 25 year contract with Georgene Novelties, Myrtle Gruelle decided that it was time for a change. It was time for a younger company that could expand more easily in the new younger market. Thus she awarded the contract for the Raggedy Ann and Andy production to the Knickerbocker Toy Company.

Myrtle kept a hand in things and in 1965 for the 50th anniversary organized birthday parties in 43 states. Knickerbocker (in honor of the 50th) gave Raggedy Ann a present. A pocket hankie for the apron of every doll.

In the 1980s the Knickerbocker Toy Company was acquired by Warner Communications, and later the toy interest was sold to the Hasbro Company.

Applause Toy Company (Hasbro Co.)
1981

Applause Toy Company started as a division of the Knickerbocker Toy Company in 1979. It produced a gift line sold through Hallmark.

Applause merged with Wallace-Barrie in the early 1980s and, since 1983 has licensed through Hasbro.

The license for these stuffed dolls is still owned by the Gruelle family.

Dolls

Raggedy Andys by Volland. 15.5". 1920-34. $1,000 each and up.

Awake and asleep Raggedy Anns by Georgene. Black outline nose. 15.5".

Raggedy Andy with button eyes and felt nose and mouth. 1950s. 24". $35-45.

1928 Raggedy Andy, stamped "Georgene Novelties Co.", cloth doll with shoe button eyes. Hand stamped, "original". 19". $150-200.

Raggedy Ann and Andy dolls with oil cloth faces, by American Novelty. 19". $200-250 each.

1918 Raggedy Ann by P.F. Volland, cloth doll with shoe button eyes and brown feet. 1915 stamped on torso. 16". Price unavailable. Rare.

Raggedy Andy by Volland.   All original. 16". $1,500 and up.

Mollye baby by Mollye Goldman. Unmarked. 1935-38. 14". Extremely rare. Price unavailable.

Oil cloth Raggedy Andy doll with stamped face. 15". $300 and up. Date unknown.

Raggedy Ann with embroidered face, by Applause. 1981. 17". $15-20.

Raggedy Andy missing button and tie, painted face, well-loved. 6.5". $6-8.

Raggedy Ann baby with down stuffing, button eyes and painted face. 9". $5-7.

Raggedy Andy "Georgene" doll sitting at his desk. 32". $150-200.

Raggedy Ann "Georgene" doll answering her phone. 32". $150-200.

Raggedy Ann and Andy "Georgene" dolls. 14". $175-225 pair.

"Georgene" Raggedy Andy with button eyes. 15 ". Stamped "Georgene Novelties Co." $70-80.

Raggedy Ann "Georgene" doll. 14". $75-90.

"Georgene" Raggedy Ann doll. 1930s. 15". Stamped "Georgene Novelties Co." $70-80.

Raggedy Ann "Georgene" doll. Black outline nose. Clothes all original, including apron and pantaloons. 19". $350 and up.

Raggedy Andy with painted face yarn hair. 32". $55-65.

Raggedy Andy "Volland" doll, shoe
button eyes. 15". 1920-34. $350 and up.

Raggedy Andy oilcloth doll with tear
drop nose. Manufacturer unknown.
14.5". $200-250.

Raggedy Anns with button eyes and
painted faces by Knickerbocker. 19".
$20-25.

Raggedy Andy with painted face, no tie, by Knickerbocker. 6.5". $6-8.

Raggedy Andy doll, no tie, button eyes, by Knickerbocker. 15". $35-40.

Raggedy Ann with button eyes by Knickerbocker. 19". $20-25.

Raggedy Ann cloth doll with button eyes by Knickerbocker. 30". $25-30.

Raggedy Ann learning doll by Playskool. Buckles, buttons, snaps and ties. 15". $15-20.

Raggedy Andy doll with button eyes, by Knickerbocker. 19". $30-35.

Early Knickerbocker Raggedy Ann and Andy. 12". $50-75 pair.

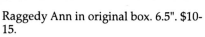

Raggedy Ann in original box. 6.5". $10-15.

Raggedy Andy "Georgene" awake and asleep doll, replaced clothes. Wears patch ever one eye. $275 and up.

Raggedy Ann and Andy. These are very old Volland dolls that were given new faces by adding material and stitching new features over the old ones. Very unusual and unique. 15.5". Rare. Price unavailable.

Raggedy Ann with no clothes, by Knickerbocker. 6.5". $5-8.

Raggedy Ann with button eyes, by Knickerbocker. 15". $10-15.

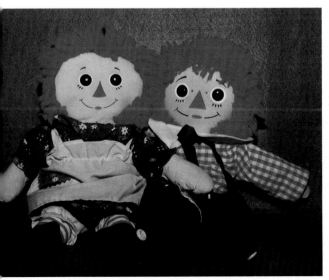

Raggedy Ann and Andy hugging dolls by Knickerbocker. Painted faces. 6". $20-30.

Raggedy Andy "Georgene" doll holding an oilcloth ball. 21.5". $350 and up.

Raggedy Andy with button eyes, sitting on a bench. 18". Knickerbocker. $20-25.

Musical Raggedy Ann by Knickerbocker. Plays "Mary Had a Little Lamb." $50-75.

Raggedy Andy with painted face, by Knickerbocker. 6.5". $10-15.

Raggedy Ann and Andy oil cloth face dolls by American Toy and Novelty Co. 1930's. 14". $500 and up.

Raggedy Andy in original box, by Knickerbocker. 6.5". $10-15.

Raggedy Ann with name on apron, by Applause. 24". $30-40.

Raggedy Ann crawling baby doll by Applause. Embroidered face, 9". $10-15.

Raggedy Andy one-piece stamped body, by Knickerbocker. 3". $8-10.

Raggedy Ann and Andy "Georgene" dolls. Top: 19" Andys. Bottom, left to right: 18" Andy, black outline nose, shoe button eyes; 18" Ann with shoe button eyes, pale narrow stripes on legs and replaced clothes;18" Andy with shoe button eyes, mint condition. $175 and up.

Beloved Belindy by Georgene. Shovel feet. 14.5". $900 and up.

Rare Beloved Belindy doll by Georgene. Replaced scarf. Marked on head. 18". 1938-1950. $1,200 and up.

Beloved Belindy by Volland, 1920s.
Original clothes. 15". $1,200 and up.

Beloved Belindy "Georgene" doll. 17.5"
with replaced scarf. Marked on hips.
1938-1950. $1,200 and up.

Beloved Belindy by Knickerbocker. 15.5"
Eyes face forward. 1965. Rare. Price
unavailable.

Beloved Belindy "Georgene" doll. Feet
move side to side. Pale chocolate color.
15". $900 and up.

Rare Beloved Belindy
"Georgene" doll with
spotted feet, markings on
back of head. 1938-1950.
Price unavailable.

Beloved Belindy doll by
Knickerbocker. 15.5".
Eyes to side. $500 and up.

Early Raggedy Ann and Andy by Volland. Ann has wooden heart. 21". 1915. Rare. Price unavailable.

Raggedy Andy "Dress Me" doll with button eyes and painted face, by Knickerbocker. 19". Teaches buttons, buckles, zips, ties and snaps. $15-20.

Raggedy Andy with no clothes, painted face, by Applause. 12". $10-15.

Raggedy Ann with button eyes, wearing a plastic Raggedy Ann necklace, by Knickerbocker (apron missing). 19". $18-22.

Raggedy Ann Mollye, redressed. 21". Made 1935-38. $700 and up.

Raggedy Andy with button eyes, stamped face. 19". $20-25.

Raggedy Ann with painted face, by Knickerbocker. 9". $10-15.

Raggedy Ann Christmas doll in original box, by Playskool. 1989. 17.5". $40-45.

Raggedy Ann with no clothes, painted face, by Knickerbocker. 6". $3-5.

Hard rubber Raggedy Andy ventriloquist dummy. 30". $100-125.

Raggedy Ann with button eyes by Knickerbocker. 16". $15-20.

Raggedy Ann sitting on a wooden tool box. 12". $20-25.

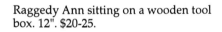

Raggedy Ann, 15". $15-20.

Bedtime Raggedy Ann doll with painted face, by Knickerbocker. 13". $10-15.

Raggedy Ann with painted face, by Hallmark Card, Inc. 6". $5-8.

Bedtime Raggedy Ann doll with painted face, by Knickerbocker. 13". $10-15.

Raggedy Ann and Andy dolls with painted faces, by Knickerbocker 5". $20-30 pair.

Raggedy Ann with painted face, by Knickerbocker. 6.5". $8-12.

Raggedy Ann with embroidered face, by Applause. "Item 8533". 9". $10-15.

Raggedy Andy with embroidered face.
"Applause". 1981. 12". $15-20.

Raggedy Andy with embroidered face,
by Applause. 12". $20-25.

Raggedy Ann doll, undressed. 15". $5-8.

Raggedy Andy sitting in his rocking chair. 18". Button eyes and buttons on his pants. $25-35.

Hugging Raggedy Ann and Andy Valentine dolls by Applause. $10-15 pair.

Homemade Raggedy Ann and Andy dolls with book. 21". $20-25 each.

Raggedy Andy painted face, by Applause. 6". $5-8.

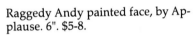

Raggedy Andy "Applause" doll sitting in old sled. 16". $18-22.

Raggedy Andy doll by Applause. 25". $30-40.

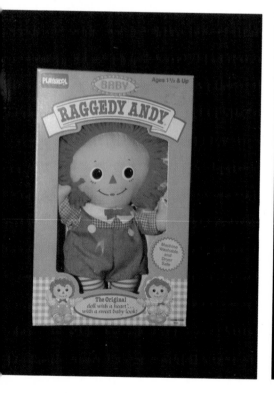

Raggedy Andy baby in the box, by Playskool. 1989. 9". $8-10.

Raggedy Ann and Andy rubber dolls in the box, by Macmillan. 1990. 7". $15-20 pair.

Raggedy Anns with button eyes, by Knickerbocker. 12". $20-25 each.

Raggedy Ann with button eyes. 7". Knickerbocker. $10-15.

Green-haired Raggedy Andy sitting on his tool box. 6". $10-12.

Handmade Dolls

Hand-made, very early Raggedy Ann doll. 26". $100-125.

Hand-made Raggedy Andy with button eyes. 15". $15-20.

Hand-made Raggedy Ann. 8.5". $10-15.

Early hand-made Raggedy Ann and Andy sitting on a bench. 16". $150-200.

Hand-made early Raggedy Ann and Andy dolls standing near the fence. 15". $175-225 each.

Early hand-made Raggedy Ann sitting with her wooden duck. 16". $75-100.

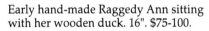

Hand-made Raggedy Ann with embroidered face and eyes. 24". $15-20.

Hand-made Raggedy Ann wearing her skates, made by Ragg-a-Muffins, Arvada, Colorado. 25". $75-100.

Hand-made Raggedy Andy with embroidered face. 14". $10-15.

Very early, hand-made Raggedy Ann doll with large paws, sitting on an old spice box. 16". $200-250.

Hand-made black Raggedy Ann. 12". $30-35.

Hand-made Raggedy Ann doll with felt eyes. 24". $10-15.

Hand-made Raggedy Sam and Sue. "Joel Martone originals". Painted faces, cotton stuffed bodies. 24". $250-300 set.

Beloved Belindy hand-made stuffed doll. Button eyes. 11". $25-30.

Hand-made Beloved Belindys.

Beloved Belindy hand-made beanbag doll with button eyes. 9". $30-35.

Hand-made black Raggedy Andy doll with button eyes and polyester suit. 25". $20-25.

Hand-made cotton Beloved Belindy with button eyes and painted face. 19". $30-35.

Hand-made Raggedy Ann and Andy cotton rag dolls with button eyes. 14". $25-30 pair.

Hand-made Raggedy Ann and Andy cotton dolls. 25". $30-40 pair.

Hand-made black Raggedy Ann and Andy cotton dolls. 19" $25-30 pair.

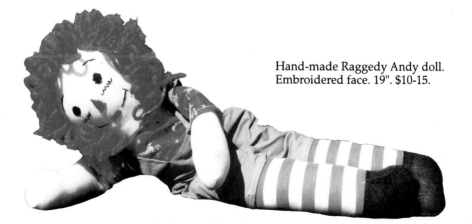

Hand-made Raggedy Andy doll. Embroidered face. 19". $10-15.

Hand-made Beloved Belindy . "Joel
Martone originals". 26". $100-125.

Hand-made Raggedy Ann with painted
eyes. 20". $10-15.

Hand-made Beloved
Belindy type doll.
Button eyes, painted
mouth. 13". $35-40.

Hand-made Raggedy Ann with button eyes and embroidered face. 24". $10-15.

Hand-made Raggedy Ann doll with embroidered face. 15". $10-15.

Hand-made Raggedy Ann, no cloths. 14". $5-10.

Hand-made Raggedy Ann and Andy
type dolls. Shelf sitters. 11". $20-30.

Raggedy Ann-type, hand-made doll,
shelf sitter. Hollow legs, painted face.
13.5". $10-15.

Hand-made Raggedy Ann doll. Embroidered face. 24". $5-10.

Hand-made Raggedy Anns with button eyes and embroidered faces. 24". $10-15 each.

Home-made Raggedy Ann and Andy dolls playing jacks. 18". $15-20 each.

Hand-made Beloved Belindy-type bean bag doll. Button eyes, painted mouth, shelf sitter. 23". $40-45.

Hand-made Raggedy Ann doll given as a gift, with enclosed letter. Button eyes. 20". $15-20.

Hand-made cotton Beloved Belindy with button eyes and painted face. 18". $30-35.

Banks

Raggedy Andy papier mache bank. 4.5".
$15-20.

Chalk carnival Raggedy Ann bank. 11".
$40-45.

Ceramic Raggedy Andy bank. 6.5". $12-
15.

Ceramic Raggedy Ann bank. 18".
"Japan". $20-25.

Ceramic Raggedy Andy bank (no hair).
3.5". $5-7.

Raggedy Ann rubber bank. 7". $30-35.

Chalk Raggedy Andy bank. 6". $8-10.

Raggedy George and Martha Washington banks. 10". "Royalty Industries". 1974. $30-35 each.

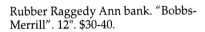

Raggedy Ann and Andy chalk banks. 8.5". "Japan". $20-30 pair.

Ceramic Raggedy Ann bank. 7". Bobbs-Merrill. $15-20.

Ceramic Raggedy Andy bank. 18". "Japan". $20-25.

Rubber Raggedy Ann bank. "Bobbs-Merrill". 12". $30-40.

Raggedy Andy hard plastic bank from Royalty Industries. 8". 1974. $30-40.

Plastic farmer Raggedy Andy bank by Royalty Co. 8". 1974. $30-35.

Ceramic Raggedy Andy bank. 7". Bobbs-Merrill. $15-20.

Chalk Raggedy Andy bank. 8". $10-15.

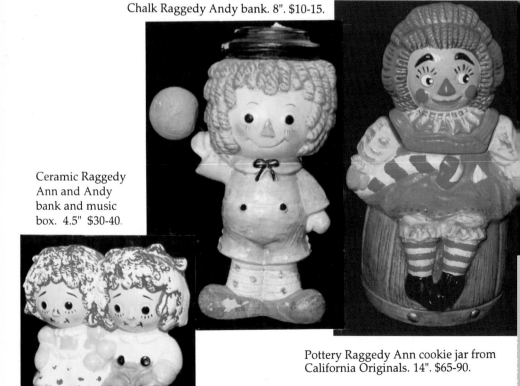

Ceramic Raggedy Ann and Andy bank and music box. 4.5" $30-40.

Pottery Raggedy Ann cookie jar from California Originals. 14". $65-90.

Figurines

Flambr Ann and Andy bisque figurines. Left to right: "Cleanliness Is"; "No ants please"; "Feeling better". $60-70 each.

Ceramic Raggedy Ann salt shaker. 3". $3-4.

Plastic Raggedy Ann figurine by Bobbs-Merrill. 4". $2-3.

Ceramic Raggedy Ann figurine. 4". $3-4.

Flambr Ann ceramic figurine with advertising plaque. $20-25.

Ceramic Raggedy Ann and Andy figurines. 4". $18-24.

Raggedy Ann chalk figurine. 6". $8-10.

Plaster Raggedy Ann and Andy door stops. 6". $15-22 pair.

Hard rubber Raggedy Ann and Andy figurines by Universal Statuary Corp. 1971. $30-40 pair.

Salt glaze dog carrying Raggedy Ann in his mouth. 3" x 5". $12-15.

Ceramic Raggedy Ann and Andy "Love one another" figurine by Bobbs-Merrill. 6". $30-35.

Ceramic baby holding Raggedy Ann doll. 9.5". $8-10.

Ceramic figurine of Raggedy Ann. 1977. 7". $15-20.

Plastic Raggedy and Andy in their paper playhouse by Macmillan. 1988. $8-12.

Raggedy Ann and Andy ceramic figurines. 4". $10-15 pair.

Hard rubber Raggedy Ann and Andy figurines by Macmillan Co. 1988. 3". $6-8 set.

Cast iron figurine by Vanity Fair. 1977. 3". $10-15.

Hard rubber Raggedy Ann and Andy figurines by Macmillan Co. 1988. 3". $5-8 pair.

Metal Raggedy Ann paper weight by Vanity. 3". 1977. $55-65.

Raggedy Ann ceramic figurine. 8". 1974. Bobbs-Merrill. $10-15.

Raggedy Andy plastic figurine by Bobbs-Merrill. 1977. 5.5". $15-20.

Papier mache Raggedy Ann and Andy Christmas ornaments. 4". $7-10.

Pair of black ceramic Raggedy Ann and Andy figurines. 4". $8-10 pair.

Plaster Raggedy Andy figure by Fritz and Floyd. 1972. $40-45.

Hard rubber Raggedy Andy figurine by Macmillan. 1988. 1.5". $3-4.

Raggedy Andy chalk figurine. "Japan". 12". $7-10.

Ceramic figurine of Raggedy Andy. 5.5". $10-15.

Hard rubber Raggedy Andy at bat. "Barrie". Made in Portugal. 3". $4-5.

Raggedy Ann ceramic figurine. 7". 1979. $8-10.

Plaster Raggedy Andy figurine. 8". $8-10.

Raggedy Ann plaster figurine. 6". "L.W. 1977". $7-10.

Raggedy Andy plaster figurine. 6". "L.W. 1977". $7-10.

Planters

Ceramic Raggedy Ann and Andy planter. 6" x 6.5". $15-20.

Ceramic Raggedy Ann and Andy planters. 5". $12-15.

Ceramic Raggedy Ann shoe planter with applied decal. 7" x 3". $35-40.

Ceramic Raggedy Andy planter. 6.5". $8-10.

Raggedy Ann and Andy plastic planters. 6.5". 1974 by Bobbs-Merrill. $20-30 pair.

Ceramic baby Raggedy Andy planter. 6.5". $12-15.

Ceramic musical Raggedy Ann planter by Bobbs-Merrill. 5.5". $25-30.

Raggedy Ann baby ceramic planter. 5". $8-10.

Ceramic Raggedy Andy planter. 6". $15-

Raggedy Andy plastic planter. 6". $8-10.

Raggedy Ann ceramic planter. 6.5".
"Japan". $8-10.

Raggedy Andy ceramic planter. 1962.
12". $15-20.

Ceramic Christmas Raggedy Ann
planter. 5.75". $12-15.

Raggedy Andy ceramic planter. 6". $10-
12.

Raggedy Andy sailor ceramic planter. 6". $10-15.

Raggedy Ann ceramic planter. 7". $10-12.

Ceramic Raggedy Andy planter. 6". $8-10.

Ceramic Raggedy Andy planter. 6.5". $8-10.

Cloth Items

Cotton Raggedy Ann baby bib. $6-10.

Felt Raggedy Ann and Andy Christmas stocking. $8-12.

Raggedy Ann and Andy cotton hand-tied baby quilt. 42" x 8". $40-45.

Raggedy Ann and Andy cotton curtains. 31" x 31". $25-30.

Pair of Raggedy Ann and Andy cotton curtains. 45". $20-25.

Raggedy Ann plastic halloween costume. 1979 by Bobbs-Merrill. $30-35.

Raggedy Ann sleeping bag. 51" tall. $70-80.

Raggedy Ann and Andy cotton laundry bag. 18" x 25". $22-28.

Set of cotton Raggedy Ann and Andy single bed, sheets. $15-20.

Raggedy Ann and Andy cotton single bedspread. $20-25.

Cotton terry cloth hand towel. $6-10.

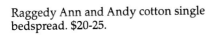

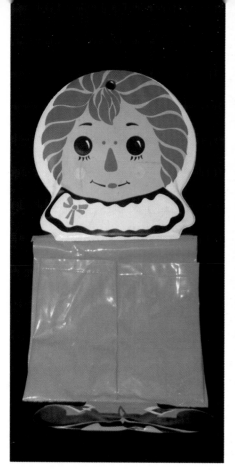

Plastic Raggedy Ann expandable child's shoe holder. 34". $20-25.

Machine-quilted, embroidered Raggedy Ann and Andy cotton baby quilt. 66" x 54". $35-40.

Plastic Raggedy Ann and Andy child's apron by Bobbs-Merrill. 1977. 18". $10-15.

Raggedy Andy cotton slippers. $5-10.

Raggedy Ann and Andy cotton curtains. 31" x 31". "Bobbs-Merrill". $20-25.

Raggedy Ann and Andy cotton pillow cases. $20-25.

Raggedy Ann and Andy cotton sheets. $15-20.

Cotton Raggedy Ann and Andy sleeping bag. $45-50.

Pillow Dolls

Cotton pillow with printed Raggedy Andy. 16". 1971 Bobbs-Merrill. $30-40.

Raggedy Andy one-piece pillow doll, printed material. 14". $20-25.

Early Raggedy Andy stuffed one piece pillow doll. The fabric is printed. 15". No markings. $35-40.

Cotton stuffed pillow with Raggedy Ann embroidered on front, patchwork background. 13" x 13". $20-25.

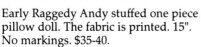

Raggedy Ann and Andy stuffed pillow dolls by Bobbs-Merrill. Printed fabric, 10". $15-20 each.

Raggedy Andy stuffed pillow doll, printed material. 24". $ 15-20.

Raggedy Ann double-sided, stamped one piece pillow doll. 6.5". $15-20.

Raggedy Ann stuffed pillow doll, printed fabric. 24". $15-20.

Raggedy Ann stuffed pillow doll, printed cotton fabric. 10". $20-30.

Unstuffed printed fabric Raggedy Ann pillow doll. 22". $15-20.

Raggedy Ann on printed fabric, cutout pillow doll with her name across her apron. 18". $10-15.

Raggedy Ann quilted pillow. 1976. 12". $50-60.

Cloth-stuffed Raggedy Ann doll. 6". $8-10.

Raggedy Ann stuffed pillow doll, printed fabric. 10.5". $15-20.

Hand-made Raggedy Ann pillow. 16". $20-25.

Cotton printed fabric, Raggedy Andy stuffed pillow doll. 18" $15-20.

Raggedy Arthur uncut pillow dog fabric. 1978. 14". $40-50.

Raggedy Ann and Andy needlepoint pillows. 11" x 13". $45-55 each.

Cotton Raggedy Andy pillow. 4" x 4". $8-10.

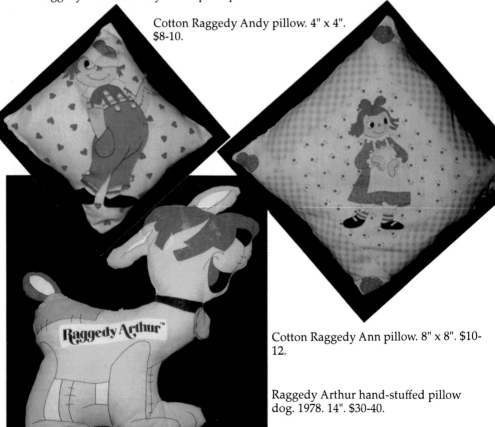

Cotton Raggedy Ann pillow. 8" x 8". $10-12.

Raggedy Arthur hand-stuffed pillow dog. 1978. 14". $30-40.

Raggedy Andy on printed fabric, cutout pillow doll, hand-stuffed. 18". $10-15.

Raggedy Ann on printed fabric, cutout pillow doll, hand-stuffed. 18". $10-15.

Raggedy Ann and Andy on printed fabric, pillow dolls. Knickerbocker. 6". $20-25 pair.

Raggedy Ann's Toy Room

Wooden Raggedy Ann and Andy baby mobile. $10-15.

Tin Raggedy Ann and Raggedy Arthur yo-yo. 2". $8-10.

Tin waste paper can with Raggedy Ann and Andy painted on front and back. Bobbs-Merrill. 1972. 13". $30-35.

Raggedy Ann and Andy plastic baby rattle ball. 13". $12-15.

Raggedy Ann and Andy pencil and pad holder. 5.5". $15-20.

Raggedy Andy pencil sharpener. 1975. "Janex Corp., Redbank, NJ". 7". $35-40.

Raggedy Ann plastic stapler. 1975. "Janex Corp., Redbank, NJ". 7". $35-45.

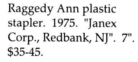

Wooden blocks with Raggedy Ann and Andy and Beloved Belindy. 1.25". $20-25.

Raggedy Ann and Andy tin sprinkling can. "Chein Playthings". 9". $20-25.

Child's high chair covered with oil cloth
Raggedy Ann fabric. $35-45.

Raggedy Ann plastic seat and metal
frame child's chair. 27". $30-40.

Raggedy Ann plastic sink by Hasbro.
1978. 12" x 21". $25-30.

Raggedy Ann plastic stove by Hasbro.
1978. 12" x 21". $25-30.

Metal "Buddy L Corp." truck. Plastic base and wheels. 11.5" x 6". $35-40.

Plastic body, rubber head, Raggedy Arthur dog figurine. 2" x 2.5". $5-7.

Plastic Raggedy Ann and animal friends by Bobbs-Merrill. 1977. $10-15.

Plastic snow dome with Raggedy Ann and Andy on a sled by Bobbs-Merrill. 1980s. $25-30.

Plastic Raggedy Andy flashlight by Bobbs-Merrill. 3.75". 1977. $20-25.

Metal Raggedy Ann and Andy crayon box. Bobbs-Merrill. 1976. 6.5" x 4.5". $18-22.

Plastic Raggedy Ann doll. 5". $5-7.

Wooden Raggedy Ann switch plate. 4" x 5". $5-8.

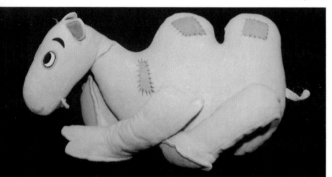

Cotton stuffed toy, the Camel with the Wrinkled Knees. 14". $100-125.

Flatsie rubber Raggedy Ann. 6". Bobbs-Merrill. $30-35.

Rubber Raggedy Andy Flatsie toy. Bobbs-Merrill. 1967. $12-15.

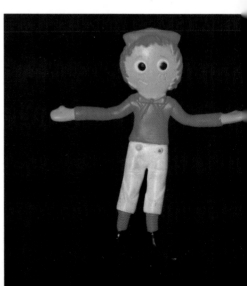

Raggedy Arthur draylon material dog. 1978. $40-50.

Metal Raggedy Ann earring holder. 6". $8-10.

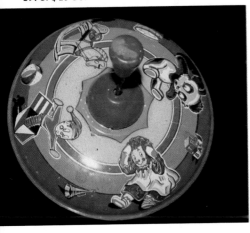

Plastic Camel with the Wrinkled Knees figure. 2.5" x 4". $7-10.

Raggedy Ann stand-up plastic comb and brush holder. 7.5". 1974. $10-15.

Tin toy spinning top with Raggedy Ann sitting on top. Chein and Co. $35-40

Plastic Camel with the Wrinkled Knees rocking horse by Bobbs-Merrill. 3.5". $10-15.

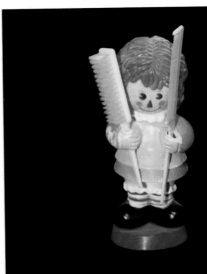

Raggedy Ann and Andy rubber squeeze toys. "Taiwan". 5.5". $25-30 pair.

Plastic Raggedy Andy socket plug by Bobbs-Merrill. 2.5". $10-12.

Plastic Raggedy Andy baby rattle. 6". Bobbs-Merrill. 1974. $5-10.

Raggedy Ann plastic and rubber baby rattle and teether. Bobbs-Merrill. 6.5". 1977. $12-15.

Pressed wood Raggedy Ann and Andy toy box by Bobbs-Merrill. 16" tall. $50-60.

Rubber Raggedy Ann bendable figurine. 1960. 2.75". $10-15.

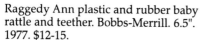

Raggedy Ann wooden doll high chair by Bobbs-Merrill. 28". $35-40.

Rubber Raggedy Ann and Andy figurine dolls. 4.5". $10-15 pair.

Plastic Raggedy Ann and Andy purse. 1988. $5-7.

Rubber Raggedy Andy squeeze toy. 5". $6-10.

Raggedy Ann plastic change purse by Hallmark. 1971. Bobbs-Merrill. 5". $5-7.

Wax Raggedy Ann and Andy candles from Price Imports, Japan. 2". $18-24.

Frosted tall glass with inserted candle, Raggedy Ann decal. 9". $10-15.

Raggedy Andy candle. 5.5". $10-15.

Raggedy Andy ceramic candle holder. 3.5". $4-6.

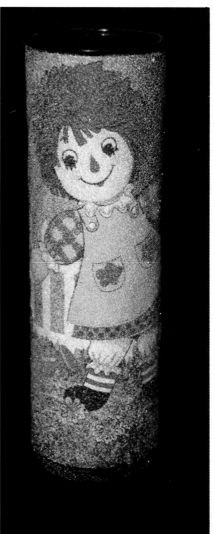

Raggedy Ann cotton hand puppet by Knickerbocker. $20-25.

Raggedy Ann marionette. Wooden head with jointed legs and plastic hands. 11". $150-175.

Raggedy Andy rubber doll by Regent Baby Products, 1973. $20-25.

Raggedy Ann styrofoam doll. "Japan". 4". $5-7.

Raggedy Ann rubber head hand puppet by Macmillan Co. 1986. 7". $10-15.

Raggedy Ann and Andy chalk book-ends. 12". 1970. Bobbs-Merrill. $30-40.

Ceramic Raggedy Ann and Andy, salt and pepper set. 3.5". $20-25 pair.

Chalk Raggedy Ann and Andy book-ends. 6.5". $15-20.

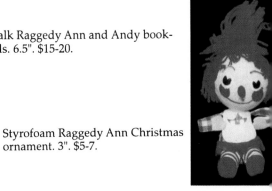

Styrofoam Raggedy Ann Christmas ornament. 3". $5-7.

Rubber Raggedy Ann doll. 5". $5-7.

Raggedy Ann and Andy chalk board. 25.5". $15-20.

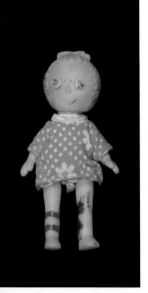

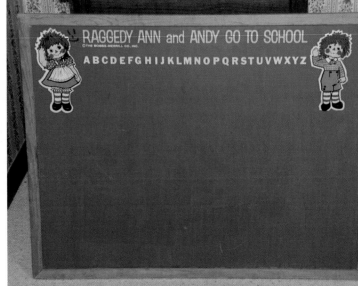

Glass Raggedy Ann and Andy Christmas ornaments. 1975. 6". $30-40 each.

Raggedy Ann plastic Christmas ornament. 3.25". $5-7.

Raggedy Andy styrofoam Christmas tree ornament. 5". $8-10.

Raggedy Andy stuffed ornament. 6.5". $10-15.

Hanging pins: Beloved Belindy & Raggedy Ann by Janice Jones, $8-10; Raggedy Ann face pin by Bobbs-Merrill, 1978, $10-12; Raggedy Ann full-figure pin from Hallmark by Bobbs-Merrill, $12-15. Peaches tin with Raggedy Ann, $8-12. Left to right: Andy figure, 3.5", Bobbs-Merrill, 1978, $6-8; Raggedy Andy pin by Bobbs-Merrill, 2.25", 1972, $8-10; Andy ornament, 3", $5-7; Raggedy Ann ornament, 4", $5-7.

Raggedy Ann and Andy styrofoam Christmas tree ornaments. 3.5". $10-15 pair.

Raggedy Ann and Andy hand-made dough ornament. 3". $3-5.

Musical/Lamps/Clocks

Raggedy Ann record player. 1974. "Vanity Fair". $50-60.

Raggedy Ann riding on her plastic winding rocking horse, which is a music box. Bobbs-Merrill. 6.25". $20-25.

Ceramic Raggedy Ann and Andy music box. "Theme from 'Love Story'". 7.5". $20-25.

Plastic Raggedy Ann and Andy transistor radio. "Philgee International". 6" x 5". $40-45.

Raggedy Ann and Andy talking plastic bank by Bobbs-Merrill. 8" x 8". 1977. Ann says, "Don't give your coins to Andy, he'll spend it on popcorn and candy." Andy says, "Don't listen to Ann, give me your money." $50-60.

Plastic Raggedy Ann and Andy record player by Bobbs-Merrill. 1975. $40-50.

Raggedy Andy plastic, music box, Santa. 8.5".

Plastic Raggedy Ann handbag by Bobbs-Merrill. 1972. $12-15.

Plastic Raggedy Ann and Andy child's change purse by Hallmark. 4". $25-30.

Wooden box with plaid cover Raggedy Ann and Andy swinging to music, music box. "Theme from 'Love Story'". 6.5" x 2.75". $25-30.

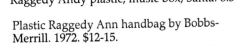

Raggedy Ann and Andy hard rubber bank. 7". $18-22.

Wooden Raggedy Ann lamp with no shade. 10.5". $15-20.

Raggedy Ann and Andy wooden musical lamp. 1974. 9" x 15". $35-40.

Wooden Raggedy Andy lamp with no shade. 10.5". $15-20.

Plastic Raggedy Ann lamp. Ann on one side Andy on the other. 1950s. 16.5". $70-80.

Raggedy Ann and Andy wooden table lamp by Bobbs-Merrill. 15". $20-30.

Raggedy Ann ceramic base for lamp. 4.5". $10-15.

Ceramic Raggedy Ann and Andy night lights. 15". $20-25.

Plaster Raggedy Ann lamp. 16". $20-25.

Wooden Raggedy Ann and Andy lamp with plaid shade. 13.75". $30-40.

Ceramic Raggedy Andy lamp. 6". $15-20.

Raggedy Ann and Andy, talking clock. "Quartz". 22". Bobbs-Merrill. $25-30.

Lamp with Raggedy Ann sitting on block with wooden base. 13.75". $20-30.

Metal Raggedy Ann and Andy alarm clock. 5". Bobbs-Merrill. 1971. $15-20.

Plastic lamp with cloth Raggedy Ann doll sitting on a block. 15". $20-30.

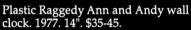

Plastic Raggedy Ann and Andy wall clock. 1977. 14". $35-45.

Plastic Raggedy Ann and Andy child's wristwatch by Bobbs-Merrill. 1975. $40-50.

Raggedy Ann watch. 1977, "Bobbs-Merrill Co." $35-40.

Plastic radio and night light by Bobbs-Merrill. 6.25". $45-50.

Plastic Raggedy Ann and Andy talking alarm clock by Janex Corp. 7". 1974. $20-25.

Pictures/Wall Hangers

Cork note board, "Have a nice day". 14" x 16". $15-20.

Raggedy Ann picture on pressed board by Lyn. 6" x 15. $5-7.

Cardboard Raggedy Ann and Andy wall plaque. 5.5". $10-15.

Raggedy Ann and Andy cardboard coat hanger. 15.5" x 18". $20-25.

Raggedy Andy sitting in high chair, cardboard wall hanger. 14". $ 10-14.

Raggedy Ann cardboard wall hanger. 14". $10-15.

Raggedy Ann and Andy and Raggedy Arthur, 3-piece cardboard wall hangers. 12".
$17-22.

Set of watercolors of Raggedy Ann done by "Edie". 1974. 11" x 15". $45-50 pair.

Raggedy Ann post letters by Hallmark. $12-15.

Raggedy Ann and Andy posters. $30-40.

Oil on canvas painting by Margery Bishop Sonnletiner. Early. 23" x 19". No price.

Raggedy Ann fiberboard cutout. Bobbs-Merrill. 1978. $20-25.

Cloth dog that is similar to the dog in Sonnletiner's painting.

Cardboard Raggedy Andy. 25". $10-15.

Cardboard Raggedy Ann. 25". $10-15.

Pictures of Raggedy Ann and Andy at play. 8.5" x 10.5". Set of 4. $15-20.

Raggedy Ann and Andy plastic mirror. 13" x 19". 1977. Bobbs-Merrill. $40-50.

Plastic wall hangers of Raggedy Ann and Andy. 1977. 13" x 8". $30-35.

Raggedy Ann and Andy plastic coat hangers. 2-piece set. 1977. Bobbs-Merrill. $40-50.

Plastic Raggedy Ann and Andy wall plaque with copper background on back of well, by Bobbs-Merrill. 13"x11". 1977. $30-40.

Raggedy Ann and Andy plastic wall hanger. 15"x16". 1977. Bobbs-Merrill. $35-40.

Raggedy Ann and Andy plastic wall hanger. 10". 1977. Bobbs-Merrill. $35-40.

Ceramic wall hanging of Raggedy Ann and Andy. 1981. 13". $20-25.

Photograph in black and white of Raggedy Ann and Andy in the window. "Ulkeke Welsh". 13" x 17". $10-15.

Print of little girl holding Raggedy Ann on wood. 5" x 7". $5-7.

Print of Raggedy Ann and Andy with baby doll. 12" x 15". $5-7.

Print of Raggedy Ann and Andy in the Raggedy school bus. 11" x 15". $35-40.

Raggedy Andy wall hanger. Plaster. 6". $8-10.

Print on felt of Raggedy Ann playing with blocks. 10" x 14". $8-10.

Hand-painted Raggedy Ann and Andy wood plaque. 10". $5-7.

Framed Raggedy Ann and Andy
gardening pictures by Lyn. 9" x 12".
$15-20 set of 3.

"Happiness is" framed picture of Raggedy Ann and Andy. 17" x 21". $15-20.

Three-dimensional framed picture of Raggedy Ann and Andy. 4.5" x 5.5". $12-15.

Wooden plaque with Raggedy Ann and Andy by Bobbs-Merrill. 8". $5-8.

Photo of Raggedy Ann by "Vince Lausen, Waterford, N.Y." 1971. 8" x 10". $15-20.

Black Raggedy Ann and Andy pictures on fiberboard. "Starco Litho of N.Y." 12" x 14". $20-25.

Wooden wall plaque poem, "My Raggedy Ann". 6" x 8". $20-25.

Framed needlepoint of Raggedy Ann and Andy sipping a soda. 17" x 17". $15-20.

Raggedy Ann and Andy pictures on fiberboard. 12" x 14" . "Stapco Co." $10-15 pair.

Paint on cloth picture of Raggedy Ann and Andy. Bobbs-Merrill. 1975. 7.5"x9.5". $10-12.

Cut outs of Raggedy Ann and Andy on cloth background. 10"x12". $8-10.

Raggedy Ann's Kitchen

Plastic Raggedy Ann and Andy bowl. 6.5". "Oneida Deluxe". 1969. $18-22.

China plate with Raggedy Ann and Andy with a chicken, from *Parents Magazine*. 7". $40-45.

Plastic Raggedy Ann and Andy bowl. 5". "Oneida Ware" 1989. $20-25.

Plastic Raggedy Ann and Andy lipped plate. "Oneida Deluxe". 1969. 8.25". $18-22.

Plastic Raggedy Ann and Andy lipped plate. "Oneida deluxe". 1969. 7.5". $20-25.

Plastic Raggedy Ann and Andy lipped plate by Oneida. 1969. 8.5". $18-22.

McDonald's giveaway Raggedy Ann and Andy Happy Meal™ with gift inside. (plastic Raggedy Ann game). 1990s. $15-20.

Raggedy Ann and Andy 4 oz. juice glass. $15-18.

Raggedy Ann and Andy drinking glasses each with different design. Bobbs-Merrill, 1977. $12-15 each.

Tin cookie can by "Bertels Can Co." 1988. "Macmillan Inc." 10" x 11.5". $10-15.

Raggedy Ann and Andy tin cookie can. 8". 15-20.

Plastic Raggedy Ann and Andy cup. 1969. "Oneida Ware". $15-18.

Raggedy Ann 12 oz. glass coffee cup. $12-16.

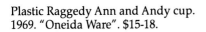

Tin cookie container. 1987. "Parco Foods Co." 10" x 3.5". $10-15.

Tin serving tray with Raggedy Ann among the toys. 1991 from Creative Creations. 12". $7-10.

Raggedy Ann and Andy plastic place mat. 14". $7-10.

Metal Raggedy Ann and Andy Thermos™ with plastic cup. 6.5". $15-20.

Metal lunch bucket with Raggedy Ann and Andy. 1973. "Aladdin Ind. Inc." $35-40.

Ceramic Raggedy Ann cookie jar by Enesco. 10.5". Small nose, green dress. $75-100.

Raggedy Andy ceramic cookie jar. 11". "Japan". $30-40.

Ceramic Raggedy Ann cookie jar by Enesco. 10.5". Large nose, blue dress. $75-100.

Raggedy Ann and Andy tin cookie can. 1973. "Cheinco Housewares". 9.5". $20-25.

Ceramic Raggedy Andy cookie jar from Japan. 10". $50-75.

Raggedy Ann and Andy tin cookie can from Cheinco. 9.5". $20-25.

Ceramic Raggedy Ann cookie jar by U.S.A #151. 12". $75-100.

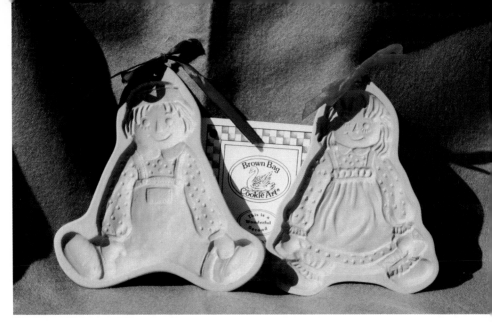

Raggedy Ann and Andy clay cookie mold by Brown Bag-Hill Design. 7". $5-7 each.

Aluminum cake mold of Raggedy Ann
by Wilton. 16". 1971. $15-20.

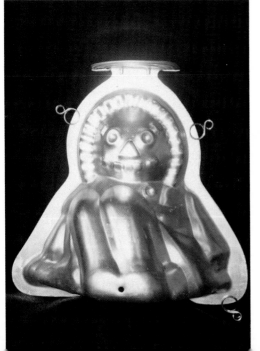

Aluminum cake mold of Raggedy Ann,
full sitting figure. 10". 1973. $15-20.

China Raggedy Ann Christmas plate. 1976. "Schmid." 7.5". $17-20.

China Raggedy Ann Christmas plate. 1977. "Schmid". 7.5". $15-18.

China Raggedy Ann Valentine's Day plate. 1978. "Schmid." 7.5". $17-20.

Raggedy Ann and Andy ceramic 1976 Mother's Day plate. "Schmid". $20-25.

Ephemera

Raggedy Ann Learns a Lesson book and record. 1978. $15-20.

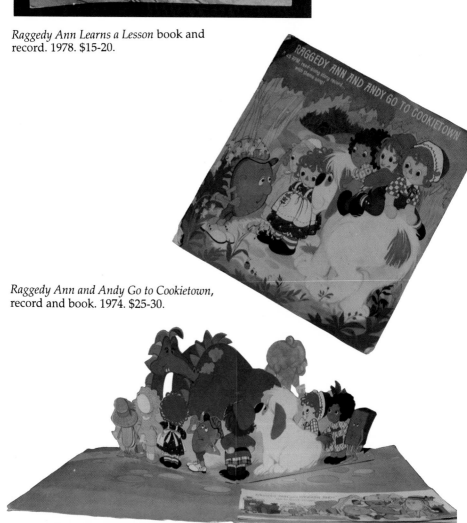

Raggedy Ann and Andy Go to Cookietown, record and book. 1974. $25-30.

Raggedy Ann and Andy Birthday Party record. 1980. $10-15.

Raggedy Ann Song Book and record. 1971. $15-20.

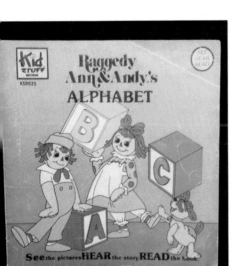

Raggedy Ann and Andy's Alphabet book and record. 1980. $15-20.

Raggedy Ann and Andy's Rainy Day Songs and Games book and record. 1980. $10-15.

Raggedy Ann and Andy "Bend and Stretch" exercise book and record, for children. Bobbs-Merrill. 1976, 1978, 1980. $25-30.

Raggedy Ann's Birthday Party book and record. 1980. $15-20.

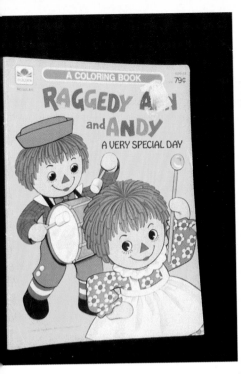

Raggedy Ann and Andy Book of Manners, record and book. 1980. $15-20.

Raggedy Ann and Andy *A Very Special Day* coloring book. 1979. $12-18.

Raggedy Ann and Andy Paper Dolls book. "Whitman Books". 1978. $20-25.

Raggedy Ann and Andy paper dolls by Bobbs-Merrill. 1966. 13". $10-15.

Raggedy Ann and Andy Circus Paper Dolls. 1974 by Whitman Books. Bobbs-Merrill. $8-10.

Raggedy Ann Colorforms. 1988. $20-14.

Cardboard Raggedy Ann paper doll. 8". $15-20.

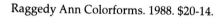

Raggedy Ann and Andy, First Doll Book. "Flip a page, change an outfit." 1969 by Bobbs-Merrill. $20-25.

Raggedy Ann "Sew and Love" by Colorforms. (Cloth doll inside.) Bobbs-Merrill. 1975.

Raggedy Ann cardboard and plastic Colorforms doll house. Bobbs-Merrill. 1974. 16" x 13". Norwood, N.J. $30-35.

Raggedy Ann and Andy "What Am I?" spinner game. Hallmark. 1978. "Bobbs-Merrill". $10-12.

Raggedy Ann and Andy game. "Rag doll fun for little people". Milton-Bradley, 1980. $18-22.

Raggedy Ann game. "A little folks game". 1974, Milton-Bradley. $15-20.

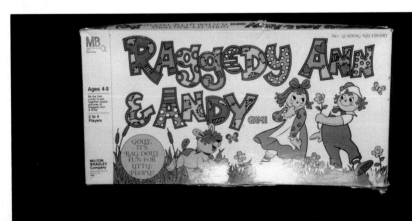

Raggedy Ann and Andy cardboard yarn pictures. Bobbs-Merrill. $15-20.

Felt Raggedy Ann with eight outfits. 1987. 12.5". $10-15.

Raggedy Ann and Andy pattern by
McCall's. 1982. $15-20.

Raggedy Ann and Andy costume
pattern. 1976. $15-20.

Roll of Raggedy Ann and Andy wall paper. $18-22.

Raggedy Ann and Andy iron on decal from McCall pattern company. 9" x 9". $10-15.

McCall's pattern for Raggedy Ann and Andy. 1982. $5-7.

Raggedy Ann and Andy puzzle. 1987. Ceaco Co. $7-10.

Raggedy Ann and Andy cardboard puzzle by Milton-Bradley. 14" x 11". 1987. $7-10.

The little rag doll with the shoe button eyes. 1950. "M. Witmark & Sons." $10-15.

Raggedy Ann and Andy paper centerpieces by Hallmark, 1974. $12-15.

Raggedy Ann and Andy Paper products: napkins by Paper Art, $5-7; stickers by Ambassador, 1988, $4-5; paper cups by Hallmark, $6-8; invitations by Paper Art, 1988; labels by Hallmark, $4-6.

Books

Wooden Willie. Dedicated to Gorden Jr. and Paul Volland. 1927. $50-60.

Raggedy Ann and Andy and the Nice Fat Policeman. Johnny Gruelle. 1960. $15-20.

Raggedy Ann and Andy's Animal Friends. Johnny Gruelle. 1974. $15-20.

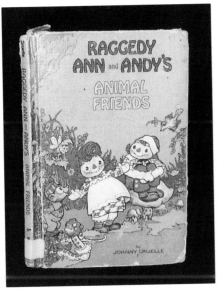

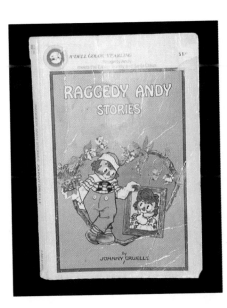

Raggedy Andy Stories. "Dell". Johnny Gruelle. 1948. $35-40.

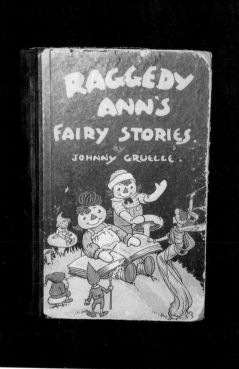

Raggedy Ann's Fairy Stories. 1928. Johnny Gruelle. $40-50.

Raggedy Ann's Lucky Pennies. 1932. $20-25.

Raggedy Ann and the Laughing Brook. 1943. Johnny Gruelle. $30-40.

Raggedy Ann's Merriest Christmas. Johnny Gruelle. 1952 $20-30.

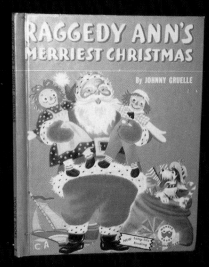

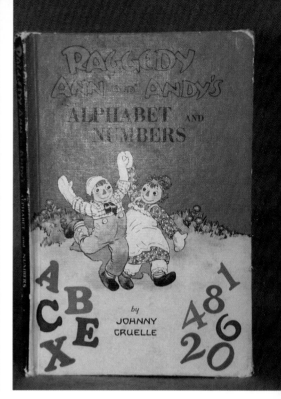

Raggedy Ann and Andy's Alphabet and Numbers. Johnny Gruelle. 1972. $15-20.

Raggedy Ann in Cookie Land. 1960. Johnny Gruelle. $18-22.

Raggedy Ann's Wishing Pebble. 1928. Johnny Gruelle. $40-50.

Raggedy Ann and the Golden Butterfly. 1932. Johnny Gruelle. $30-35.

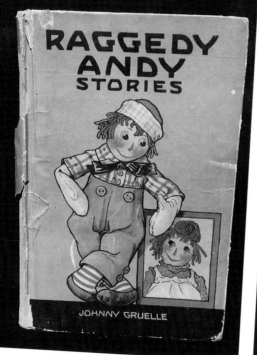

Raggedy Andy Stories. 1920. $40-50.

Raggedy Ann and the Golden Butterfly. 1940. $30-40.

Raggedy Ann and the Hobby Horse. 1961. Johnny Gruelle. $15-20.

Marcella, a Raggedy Ann Story. 1929. $40-45.

Raggedy Ann and the Wonderful Witch.
1961. Johnny Gruelle. $15-20.

Raggedy Ann and the Happy Meadow.
1961. Johnny Gruelle. $15-20.

Raggedy Ann in the Deep Deep Woods.
1930. Johnny Gruelle. $30-35.

Raggedy Ann and Betsy Bonnet String.
1943. $25-30.

Raggedy Ann and Andy and the Camel with the Wrinkled Knees. Original box. 1924. $60-70.

Raggedy Ann in the Magic Book. Illustrated by Worth Gruelle. 1939. $40-50.

Little Sunny Stories by Johnny Gruelle. 1919. $50-60.

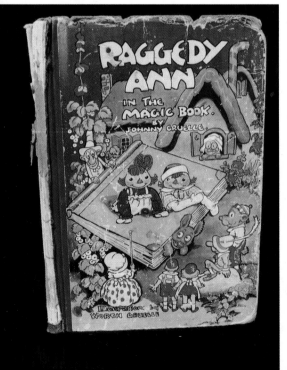

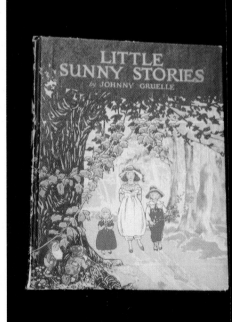

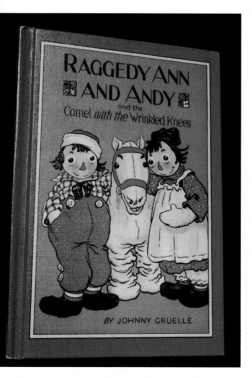

Raggedy Ann and Andy and the Camel
with the Wrinkled Knees. 1960. $20-25.

Raggedy Ann Stories, by Dell Pub. Co.
1977. 10-15.

Raggedy Andy's Surprise. "Wonder
Book". 1953. $15-20.

Raggedy Ann's Wishing Pebble. 1960. $10-
15.

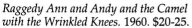

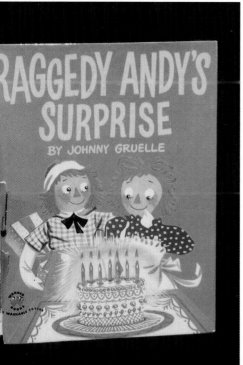

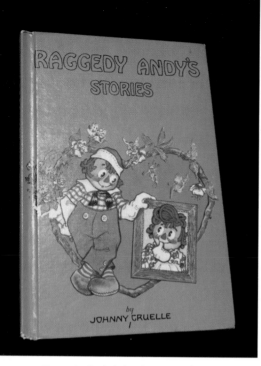

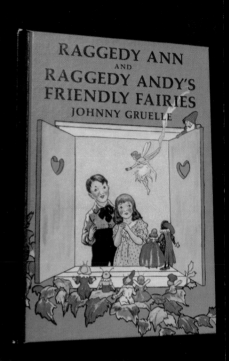

Raggedy Andy's Stories. 1960. $10-15.

Raggedy Ann and Raggedy Andy's Friendly Fairies. 1960. $15-20.

Raggedy Ann Stories. 1960. $12-18.

Raggedy Ann and the Golden Ring. 1961. $10-15.

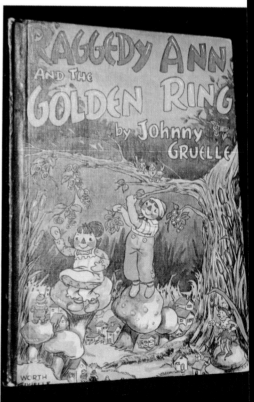

Beloved Belindy. 1960. $45-50.

Baby's First Book, Play with Raggedy Ann and Andy. Washable by Wonder books. $8-10.

The Raggedy Ann Book. "Golden Shape Book". 1969. $8-10.

The Raggedy Ann Book by Janet Fulton. 1969. $8-10.

Tell a Tale Book, *Raggedy Ann and the Tagalong Present*. 1971. $10-15.

Raggedy Ann's Tea Party. "Wonder Books". 1954. $20-25.

Little Golden Book, *Raggedy Ann and Andy and the Rainy Day Circus*. 1980. $5-10.

Raggedy Ann Stories to Read Aloud. 1960. Wonder Books. $8-10.

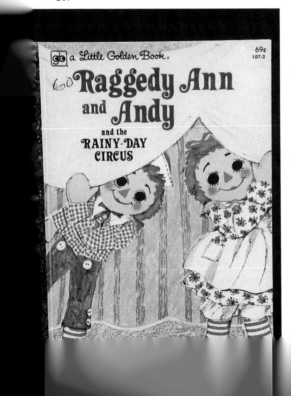

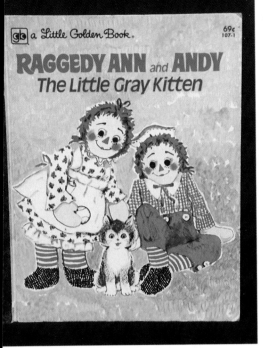

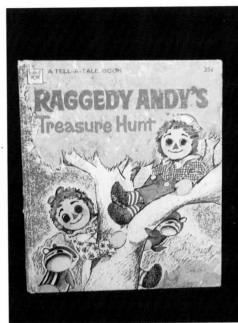

Little Golden Book, *Raggedy Ann and the Little Gray Kitten.* 1979. $15-20.

Raggedy Andy's Treasure Hunt. "Wonder Books". 1973. $15-20.

Tell a Tale, *Raggedy Ann's Cooking School.* 1974. $10-15.

Raggedy Ann's Tea Party. "Wonder Book". $15-20.

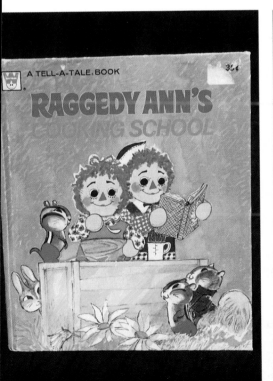

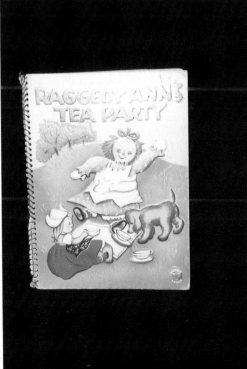

Raggedy Ann's Secret. 1959, by Wonder Books. $18-22.

Golden Book, *Raggedy Andy, I Can Do It, You Can Do It Book.* 1973. $10-15.

Little Golden Book, *Raggedy Ann and Andy Help Santa Claus.* 1972. $10-15.

Little Golden Book, *Raggedy Ann and the Cookie Snatcher.* 1972. $10-15.

This is *Raggedy Ann. Baby's First Book,*
wipes clean. "Golden".1970. $10-15.

Raggedy Ann and the Cookie Snatcher.
"Little Golden Books". 1972. $15-20.

Raggedy Ann's Favorite Things. "Golden
Cloth Book." 1972. $15-20.

The Raggedy Ann and Andy Book. 1982.
$5-10.

Golden Book, *Raggedy Ann, A Thank You, Please, and I Love You Book*. 1978. $5-10.

Raggedy Ann's Picture-Perfect Christmas. Gail Herman. 1988. $5-10.

Raggedy Ann's Adventure. 1947. $20-25.

Raggedy Ann's Mystery. 1947. Drawings by Ethel Hays. $20-25.

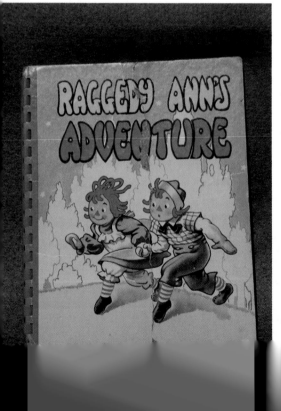

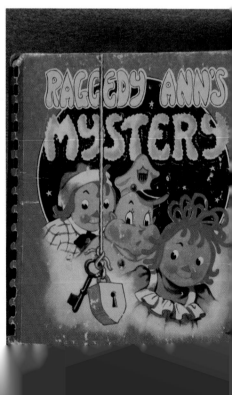

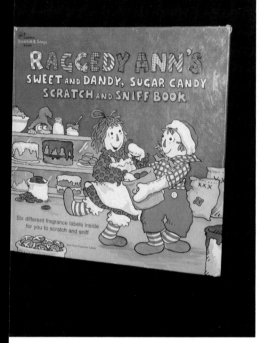

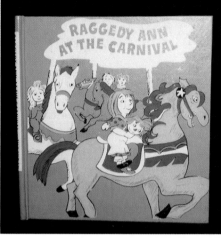

Raggedy Ann at the Carnival. 1977. $15-20.

Raggedy Ann's Sweet and Dandy, Sugar Candy, Scratch and Sniff Book. 1976. $15-20.

Raggedy Ann and Andy at the Zoo book of soft washable plastic. 1974. Bobbs-Merrill. $10-14.

Raggedy Granny Stories. 1977. Doris Thorner Salzberg. $20-25.

Raggedy Ann and Andy and the Magic Wishing Pebble. Cathy East Dubowski. 1987. $10-15.

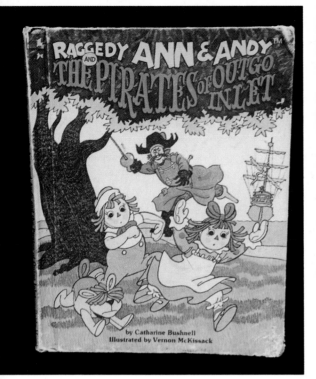

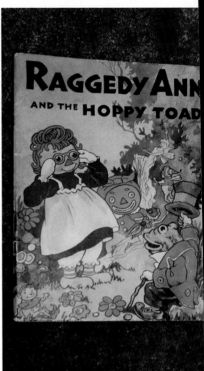

Raggedy Ann and Andy and the Pirates of Outgo Inlet. "Weekly Reader". 1980. $5-10.

Raggedy Ann and the Hoppy Toad story book. "Sherman Bagg, 1977." $15-20.

Where Are You Raggedy Andy? Photographs by Anita & Steve Shevett. 1987. $5-10.

Raggedy Ann and Andy's Grow and Learn Library, Vol. 1, Sunny Bunny Comes Home. 1988. $20-25.

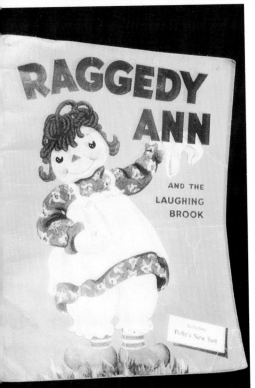

Raggedy Ann and the Laughing Brook.
"Perks Pub. Co." 1946. $25-35.

Raggedy Ann Helps Grandpa Hoppergrass
book. "Perks Pub. Co." 1946. $20-25.

Raggedy Ann in the Garden. "Perks Pub.
Co." 1946. $25-35.

Weekly Reader, *Raggedy Ann and Andy
in the Tunnel of Lost Toys.* 1980. $5-10.

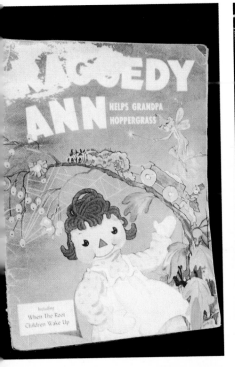

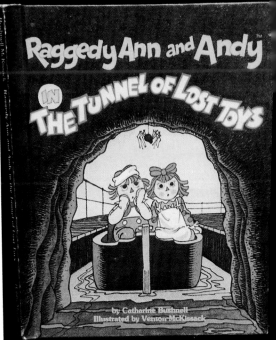

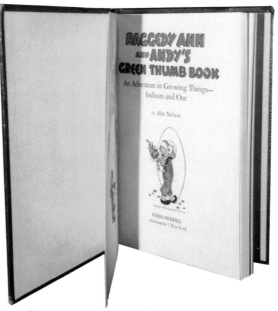

Raggedy Ann ana Andy's Green Thumb Book. 1975. Bobbs-Merrill. $18-22.

Comic book. *Raggedy Ann and Andy in the Hair Raising Adventure of Peterkin Pottle.* Vol. 1. #33. 1949. $30-40.

Comic book. *Raggedy Ann and Andy.* Vol.1 #2. 1946. $30-40.

Raggedy Ann and Andy Giant Treasury. Johnny Gruelle. 1984. $30-40.

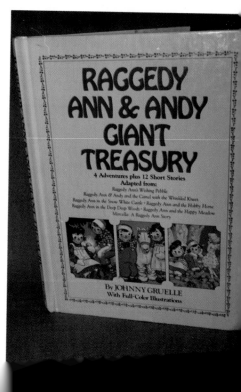

Early *Raggedy Ann and Andy* D.C comics. Left to right: Dec. 1947; Aug. 1947; Feb. 1948. $30-40.

Little Treasury of Raggedy Ann and Andy miniature story books. Set of 6. 1984. $10-15.

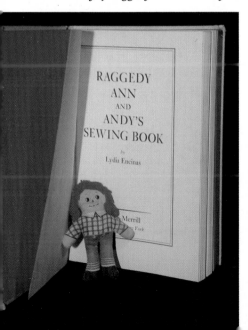

Raggedy Ann and Andy's Sewing Book. No picture on green cloth cover book. 1st. printing. 1977. $15-20.

Little Treasury of Fairy Tales. Included is "Raggedy Andy's Pillow Fight." Ill. by Gerry Embleton.1984. $8-10.

Handmade Miscellaneous

Hand-made "Joel Martone originals". Composition Raggedy Ann pin. 2.5". $15-20.

Hand-made Raggedy Ann felt head pin. 3". $3-4.

Composition Raggedy Ann pin. 2.5". $5-6.

Fur Raggedy Ann and Andy mice. 3". $5-7.

Hand-made cardboard Raggedy Ann by "Joel Martone". 42". $30-40.

Hand-made cardboard Raggedy Ann by "Joel Martone". 35". $30-40.

New wooden Raggedy Ann figure. 3".
$5-7.

Resin Ann and Andy ornaments. 3". $8-
10.

Wood pencil with Raggedy Ann head.
8". $5-7.

Hand-painted Raggedy Ann mail box
by Annette Mahanti. $75-100.

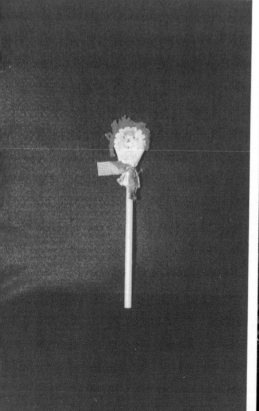